Sustainable Electricity

Heather Rising

Contents

Sustainable Electricity

What Is Electricity?

Electricity is a form of energy that is found naturally in lightning bolts and **static charges**. Over 300 years ago, scientists discovered how to make their own electricity, and today, electricity can run machines, cool houses and power lights. Most electricity is made by burning **fossil fuels** such as coal or gas, which releases carbon dioxide into the air and contributes to **climate change**. Once fossil fuels are burned, they are used up, so they are not able to be used again.

Electricity can be found in nature, such as when lightning strikes during a storm.

The human body makes electricity to power muscles, nerves and the brain. Some animals, such as electric eels, produce electricity to stun their prey.

When fossil fuels are burned, they release carbon dioxide and pollution.

Every year, the world's demand for electricity grows. If we make electricity through **sustainable** methods, such as by using wind or solar energy, it will be available far into the future and there will be little harm to the environment.

Wind can be used to make electricity without causing significant harm to nature.

Discovering Electricity

The ancient Greeks experimented with making electricity and discovered that by rubbing two materials together, they could create a very small static charge. Static charges can be created at home by actions such as pulling clothes from a dryer or shuffling your socks on a carpet.

In the late 1800s, scientists discovered that electricity was the movement, or flow, of **electrons** between two substances. Electrons are found in atoms, which are the tiny particles that everything in the world is made up of. Scientists learnt they could move electrons by turning magnets around a copper wire. They built a device called a **generator**, which could create a steady movement of electrons.

Parts of an Atom

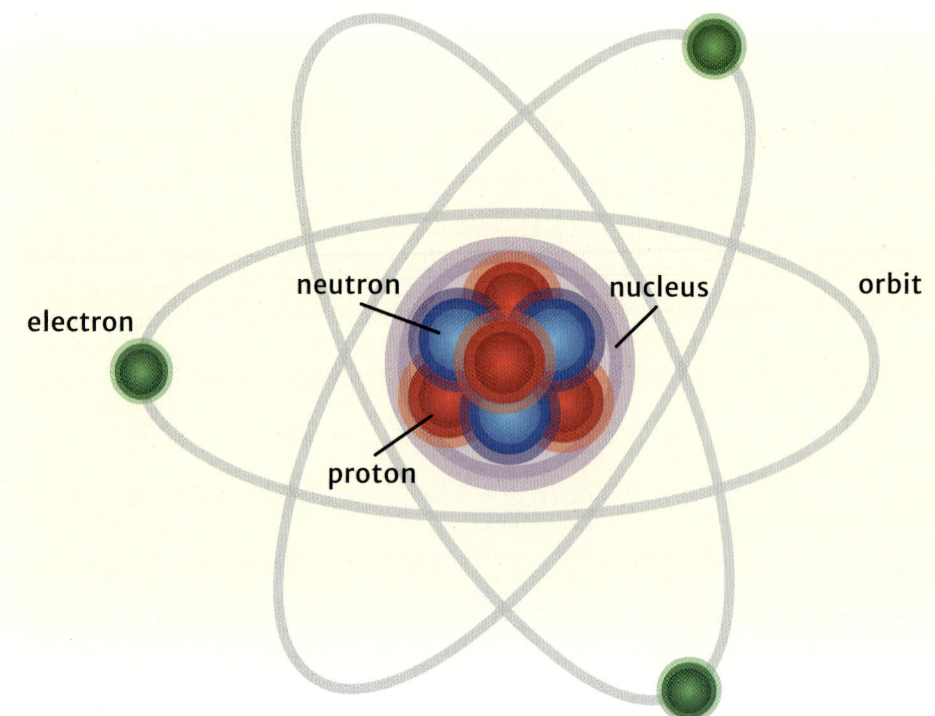

The flow of electrons along a copper wire creates electricity, or an electric **current**. Electricity flows easily through copper, as well as some other materials, which are called "conductors". Different materials, such as wood, don't allow electricity to flow through them easily. These are called "insulators".

Most electrical wires have copper inside them to help electricity flow easily from one place to another.

Some artists use electricity to make art. Introducing electrical charges to wood creates interesting patterns, because electricity cannot flow well through it.

Making Electricity with Chemicals

Scientists continued to do experiments to find other materials that could move electrons easily. In 1800, a scientist named Alessandro Volta discovered that, in the right conditions, copper and zinc placed in salt water could make an electric current. He then invented a container that made a steady flow of electricity, which could be stored until needed. This container was later called a battery.

Battery design improved over time, and soon batteries could be reset, or recharged. Electricity could flow from them again and again. This meant that batteries could be reused, which made them more affordable. Today, many of our devices are powered by rechargeable batteries.

This model of Alessandro Volta's battery shows how copper and zinc discs were stacked in a container that could be filled with salt water.

Modern batteries, such as this car battery, can be recharged multiple times.

Scientists are continuing to search for ways to improve batteries, with the goal of making a greater amount of electricity that is forever rechargeable. Batteries or other similar containers may be a sustainable way to make electricity in the future.

Many batteries can be linked to create lots of electricity, like this system that acts as a power backup in a factory.

Rechargeable batteries made electrical power affordable and led to the invention of electrical machines. By 1897, electric taxis running on batteries were being driven on roads in the UK and the USA.

an electric taxi in London, England

Demand for Electricity

Electric Generators

The invention of electrical machines and efficient light bulbs in the early twentieth century created a bigger demand for electrical energy. Huge generators with fan blades, called "turbines", were designed to make large amounts of electricity. Some early turbines used the power of falling water to spin the blades, such as the first generator built in Niagara Falls, USA, in 1882.

This power station on the Canadian side of Niagara Falls has been made into a museum.

an example of an early toaster from the 1920s that used electricity

Many new electrical appliances were invented in the late 1800s. Things like the electric toaster, coffee maker and hairdryer were all invented before most homes were even wired for electricity.

Around the same time, electricity generating stations with multiple turbines were built in large cities, using steam to move the turbine blades. These stations burned coal to heat water until it boiled and made steam. However, electricity generated by fossil fuels is not sustainable, because burning these fuels releases gases that cause climate change. Removing fossil fuels from the ground can also contaminate water supplies and destroy habitats. Accidental oil leaks can damage both land and ocean ecosystems.

Penguins are some of the animals that can be affected by oil spills, which can happen when oil is drilled or transported.

Making Electricity Sustainable

In many countries around the world, scientists are changing how **power plants** work so that they produce sustainable electricity. They can do this by finding **renewable** power sources to turn the turbines, or by developing an alternative way to make electricity that does not harm the environment.

Hydroelectric Power

When water is used to spin a turbine and make electricity, it is called hydroelectric power. This process doesn't produce any harmful gases and the water can be reused, so it is a sustainable way to make electricity. Hydroelectric plants are built near waterfalls and rivers. Some rivers are **dammed** to create a steady, powerful stream of water to spin the turbines. Today, almost 16 per cent of the world's electricity is produced by hydroelectric sources.

The Hoover Dam is a large hydroelectric plant in Nevada, USA.

However, hydroelectric power has limitations. For example, a plant must have access to fast-moving water. The amount of water in rivers can also be affected by the level of rainfall, with some rivers drying up and others bursting through the dams. In some cases, human settlements have been destroyed by flooding caused by dams.

Damming rivers to hold water can sometimes damage the surrounding habitat and wildlife. Interrupting the natural flow of water in rivers prevents fish from moving freely, which can disrupt their life cycles.

Nuclear Power

In nuclear power plants, turbines are spun by steam created from heat made in a nuclear **reactor**. When the heat is later released, it is free of harmful gases, and the materials used to make nuclear power are plentiful on Earth. This means that it is a sustainable way to make electricity. However, there are some major drawbacks. The waste material created in a reactor can remain **radioactive** for up to hundreds of thousands of years, which means it is extremely harmful to living things. The waste must be stored safely, which can be challenging and costly. If radioactive material was accidentally released, it would seriously harm the surrounding ecosystem.

Nuclear reactors, like these ones in a nuclear power plant in Japan, have a distinctive dome shape.

The first nuclear power plants came into operation in the 1950s, and there are now over 400 nuclear power plants around the world. They produce around 10 per cent of the world's electricity needs.

In 2011, a powerful earthquake caused a tsunami with 10-metre-high waves, which hit and damaged the Fukushima Daiichi nuclear power plant in Japan. Several reactors suffered a **meltdown**, which caused radioactive material to leak into the air and the ocean. It is considered one of the worst nuclear accidents in history.

A rescue team wearing protective suits helps clean up after the Fukushima accident.

Wind Power

Wind is plentiful throughout the world, so it is a sustainable way to generate electricity. Wind has been used for thousands of years to turn windmills, which were then used for pumping water or grinding grain.

This old windmill in Spain was used to grind grain that grew in the area.

The first known wind turbine was built in 1887 by James Blyth, to make electricity for his own home. By the early twentieth century, wind turbines were mainly used in rural areas, where farms and houses were far from power lines.

By the 1970s, improvements in technology made it easier for lots of wind turbines to be made and used. However, wind turbines are typically about the height of a 25-storey building and must be placed in open areas, such as on farmland, along highways or out in the ocean. Wind turbines also need to be built in areas that receive regular, steady wind. They are often built in groups and placed facing different directions to reduce periods when no electricity is being made.

Since the early 2000s, the amount of electricity made from wind power has grown rapidly. In 2022, it accounted for just over 5 per cent of the world's electricity production.

An average modern wind turbine can make enough electricity in under one hour to power a home for about a month. In some years, Denmark generates as much as 47 per cent of its electricity from wind turbines.

Farmland is a good place for wind turbines, as there's enough open space for them to catch lots of wind.

These wind turbines in Copenhagen, Denmark, were built on the water so they would have plenty of space.

Tidal Power

In order to find new sustainable ways to make electricity, researchers began to examine the power of tides. The ocean is in constant motion due to the pull of the Sun and Moon's gravity, making tides a dependable and constant source of energy to spin turbines.

The first commercial tidal power plant opened in France in 1966, and it is still in operation today. It has 24 turbines, and they can spin in both directions so that the blades can move when the tide goes in and out. The plant generates 240 megawatts of electricity every day. A megawatt is equal to one million **watts**. The plant makes enough electricity to power 130 000 homes.

The Rance Tidal Power Station in France was built across a river close to the ocean so that the turbines could be spun by the tides.

While tidal power can make sustainable electricity, there are still some concerns about the impact the turbines can have on sea life. To reduce damage to living things in the ocean, engineers are testing turbines that float on top of the waves.

The Sihwa Lake Tidal Power Station in South Korea has been the world's biggest producer of tidal power since 2011. The turbines were built using an artificial sea wall, and they spin when the tide comes in. The plant produces 254 megawatts of power every day.

In Orkney, Scotland, several types of tidal turbines have been installed to capture electricity from the sea.

Geothermal Power

Below ground, there is an endless source of heat from Earth's core. Heat is brought up from the core by the movement of magma, or hot liquid rock. In some places, this heat is very close to the surface, particularly where there are active volcanoes. The heat from Earth's core can be captured to make electricity. The first **geothermal** plant, built in Italy, was able to use this electricity in 1913.

Sometimes, geothermal energy causes hot water to burst from under the ground.

The Larderello plant in Italy, the first geothermal plant, is now one of many in the area that is able to use geothermal energy.

The largest power-producing geothermal plant is The Geysers Geothermal Complex in California, USA. It is spread over 100 square kilometres and generates enough electricity to power one million homes.

To use geothermal energy, a hole is drilled into an underground supply of heated water. The water is hot enough to turn into steam, which is then piped through a plant and used to spin the turbines. As the steam cools, it returns to its liquid form and is pumped back into the underground well. The Earth constantly reheats this water, so it is a sustainable way to make electricity.

The heat below ground is not easy to access in many areas, but geothermal plants can be built in any place that has volcanic activity, such as Italy, Iceland and the Philippines.

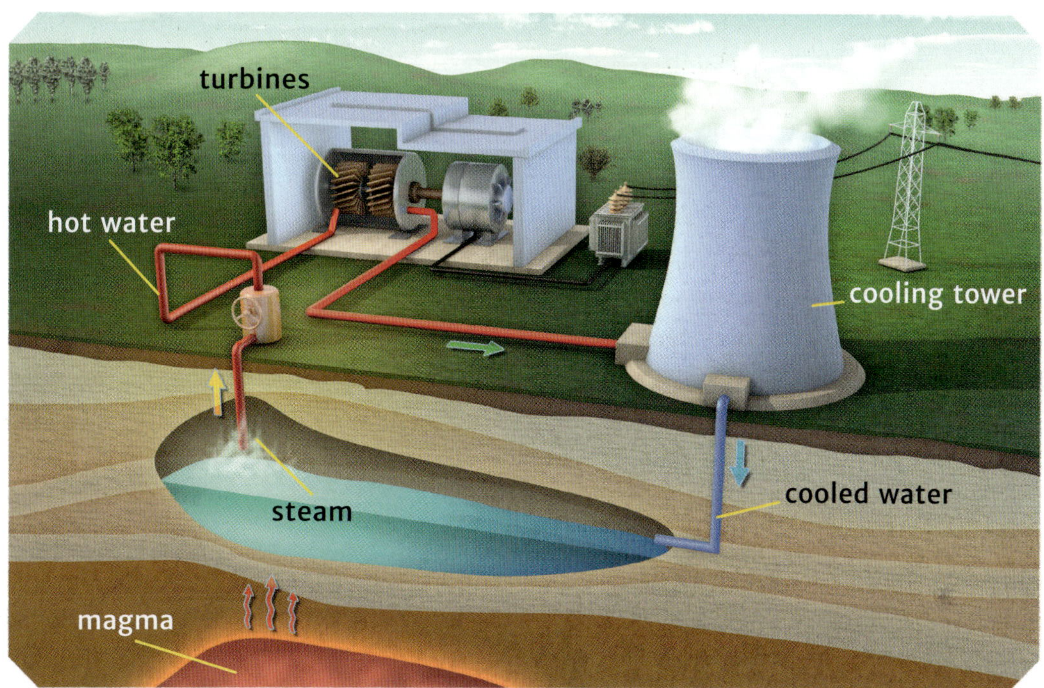

This diagram shows how steam created by heat below the ground can be used to spin a turbine, before the cooled water is returned underground.

Solar Power

Another source of energy we have plenty of is the Sun. Experiments to change light and heat energy from the Sun into electricity began in the 1800s. But it wasn't until the 1950s that scientists discovered that silicon was excellent at creating an electric current when used in a **solar cell** and exposed to sunlight.

Silicon is a hard substance found in almost all rocks.

In 1958, the USA launched the first satellite powered with solar cells into space. In 1976, the first solar-powered calculator became available, but it took another four decades for solar technology to become cheap enough to use in other products. Solar cells are now found in watches, battery chargers and on the roofs of houses and buildings.

The International Space Station is a satellite that is fitted with large solar panels on its sides so that it never runs out of electricity.

Solar cells provide all the electricity for the International Space Station and Mars rovers, as well as for new satellites and space telescopes.

Solar power contributes about 3 per cent of the world's electricity production, but this number is increasing quickly. The largest solar farm, the Bhadla Solar Park, opened in India in 2018, and it produces 2250 megawatts a day.

Large solar farms, like this one in Egypt, can be built in areas that receive a lot of sunshine.

However, there are limitations to solar power. A plant can only produce electricity during daylight hours and on sunny days. Large solar farms also take up huge areas of land and can change the habitat for local wildlife.

Scientists are continuing to improve the design of solar power plants in order to reduce their impact on the environment. Building solar panels in areas that are already clear of trees is one way. Solar farms can also now float on lakes, reservoirs and even in water tanks. The panels receive direct sunlight while providing shade for the water below. This can reduce harmful bacteria growth and prevent too much **evaporation**. There are also projects experimenting with floating solar farms on the ocean.

This solar farm in Taiwan has been built on a lake in order to collect lots of sunlight.

Solar cells can now be made much thinner, and they can even be 3D-printed. These types of solar panels line some bike paths and highways and are used to power street lights and road signs. They can also be added to the tops of vehicles, where they can be used to charge the vehicle's battery when it is in sunshine.

Solar panels can power road signs, such as the ones on this corner in Thailand.

The Future of Sustainable Electricity

The future of sustainable electricity may lie in technology that is in the early stages of development.

Chemical Fuel Cells

Similar to experiments with early batteries, scientists are now experimenting with chemical reactions that can generate large amounts of electricity. These reactions take place in a type of container called a fuel cell. There are several kinds being tested, but the most common uses hydrogen gas, which reacts with oxygen to produce electricity. Since hydrogen is abundant on Earth, electricity can be made sustainably this way. However, hydrogen can catch fire easily, so the fuel cells must be expertly handled.

Hydrogen fuel cells can even be used in vehicles, like this one that powers a truck engine.

Microbial Fuel Cells

Scientists are also working on microbial fuel cells (MFCs). These use **microorganisms** such as bacteria, algae and some fungi to produce electricity. When a living thing changes food into energy for itself, it can produce electrons. These electrons can be collected to create a current strong enough to recharge batteries. MFCs have been powering environmental sensors on the ocean floor since 2016, which can measure either the water temperature or earthquake activity.

Researchers have also been testing plant microbial fuel cells (PMFCs). PMFCs collect electrons from the bacteria in soil that feed on the waste from growing plants. PMFCs are already being used at a university in the Netherlands, and engineers are looking for ways these PMFCs could be placed under farms and garden lawns to collect electrons on a large scale.

A scientist demonstrates how PMFCs can be used to power a toy car.

Capturing Movement

A company in the Netherlands has found a way to capture the energy from human movement and turn it into electricity. In 2008, they invented a dance floor that makes electricity from the movement of dancing. Several countries now have dance clubs that use dancing to power the lights and music.

In addition, many electric bicycles are now self-charging. The rider can generate electricity for the battery when they pedal. There is also exercise equipment that generates electricity when it is used. As users get in shape, they can make electricity and send it right into the grid that powers their home.

An electric dance floor was set up in London for people to learn how to Irish dance and power a TV screen with their steps at the same time.

Scientists are also creating new materials at an atomic scale, called **nanotechnology**. They have invented substances that generate electricity when water evaporates from the surface. Eventually, these substances could be used to coat outdoor objects, to generate electricity both when rain hits them, and again when the water evaporates. Scientists have also invented materials that generate static electricity when they are rubbed or moved. Perhaps one day, people's clothing could generate electricity as they move or could even recharge the battery in a phone.

A scientist works on creating new substances by using nanotechnology.

Scientists are testing micro turbines that can generate electricity when blood flows through a vein. In the future, these tiny turbines might power small medical devices in people's bodies.

The demand for electricity is only increasing. Scientists have spent the last few hundred years learning how to create electricity and finding sustainable ways to make it. Choosing to power our homes, cars and industries with sustainable electricity will help keep our planet safe for people in the future.

Experiments to Create Static Electricity

There are ways to experiment with electrons at home and make your own static electricity. By rubbing two objects together, electrons can be moved from one material to another. When this happens, the objects can become “charged”. This means that the atoms in the object have either fewer electrons (a positive charge) or more electrons (a negative charge) than usual. Having a positive or negative charge can cause an object to attract, or move closer, an object that is not charged.

Goal

To create static electricity in two experiments

Materials

You will need:

- 2 balloons

- a tablespoon

- salt

- ground black pepper

- a small plate

- a woollen item of clothing, such as a scarf or sock

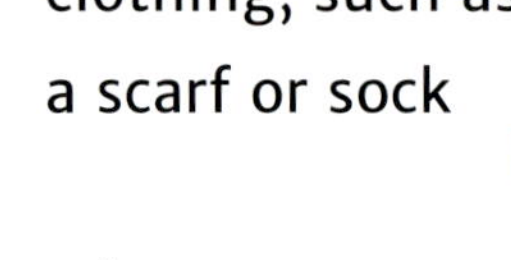

- a tap

Steps

Experiment 1

1. Blow up and tie the first balloon.
2. Measure out a tablespoon of salt and a tablespoon of pepper, and place them on the plate. Mix them together.
3. Move one balloon over the salt and pepper, to check that the salt and pepper don't move.
4. Rub the balloon back and forth on your hair. The balloon is now negatively charged with extra electrons taken from your hair.
5. Move the balloon over the plate of salt and pepper again. The pepper grains, which are lighter than the salt, will be attracted to the negative electrons on the balloon.

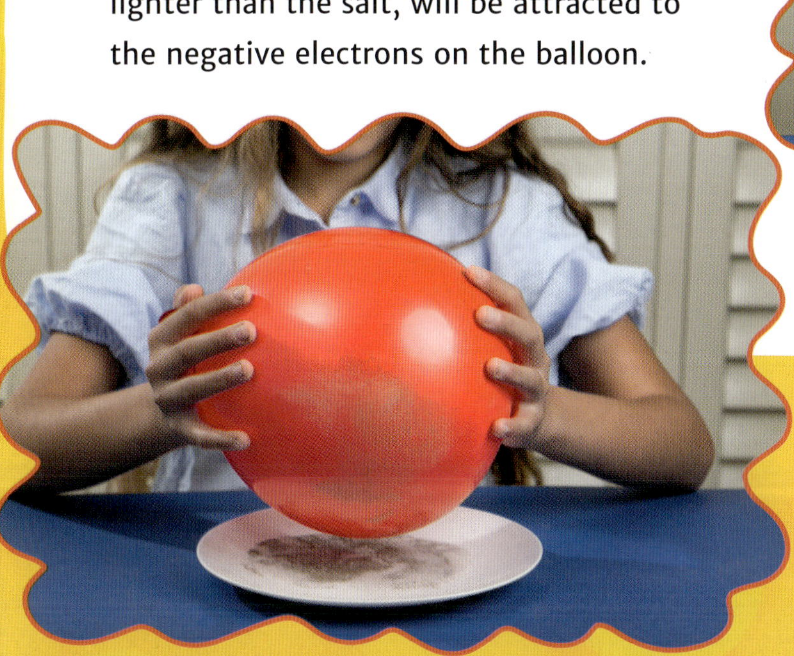

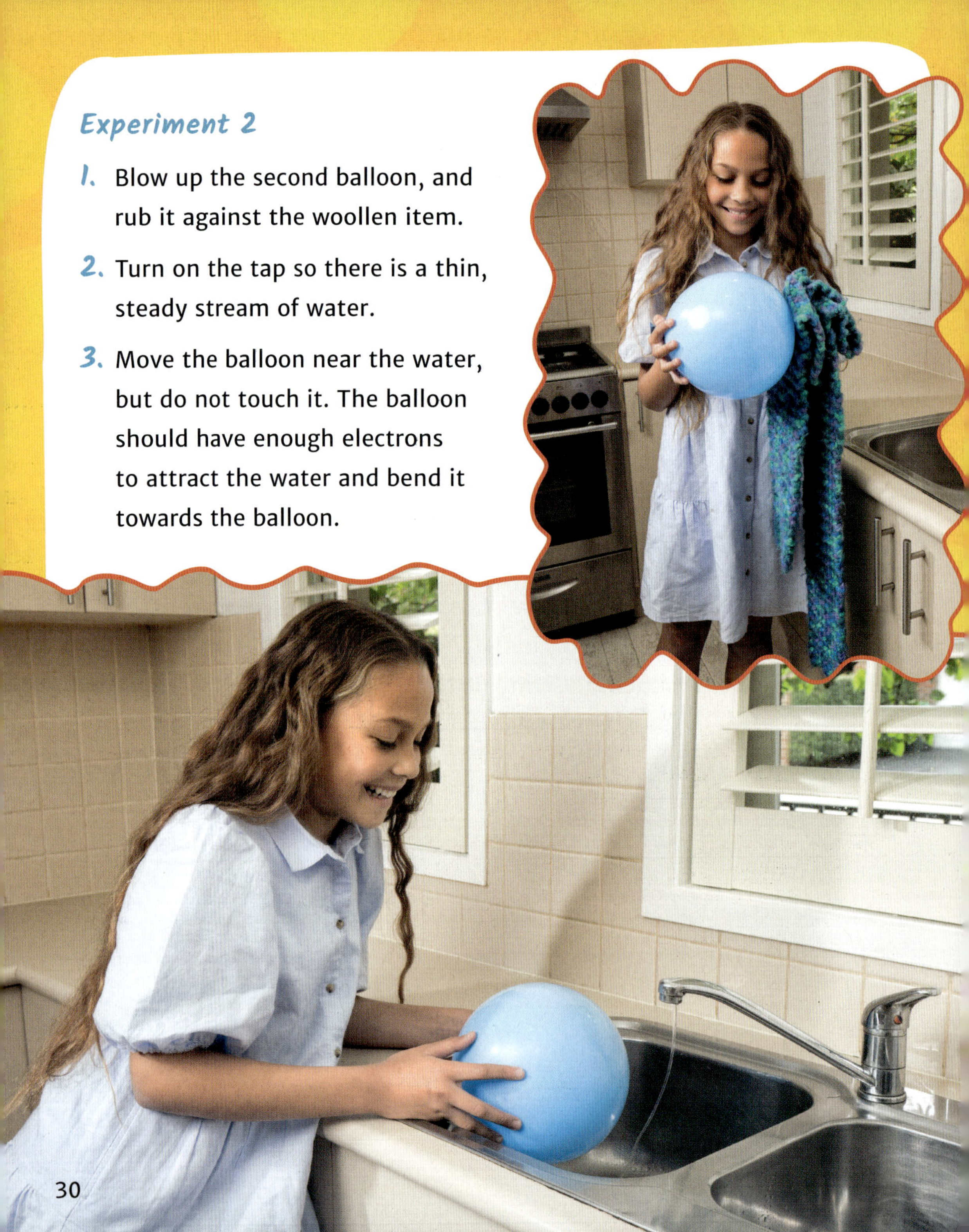

Experiment 2

1. Blow up the second balloon, and rub it against the woollen item.
2. Turn on the tap so there is a thin, steady stream of water.
3. Move the balloon near the water, but do not touch it. The balloon should have enough electrons to attract the water and bend it towards the balloon.

Glossary

climate change (*noun*)	a global change in weather patterns
current (*noun*)	a flow of electricity
dammed (*verb*)	partly blocked to cause water to flow in a narrower, more powerful stream
electrons (*noun*)	small parts of atoms that have a negative charge
evaporation (*noun*)	when a liquid changes into a gas
fossil fuels (*noun*)	fuels, such as coal, oil and petrol, which are made from dead plants and animals that were buried millions of years ago
generator (*noun*)	a machine that makes electricity
geothermal (*adjective*)	using heat from under the ground
meltdown (*noun*)	an accident where harmful nuclear materials are released
microorganisms (*noun*)	tiny living things
nanotechnology (*noun*)	an area of science that deals with structures that are very tiny
power plants (*noun*)	buildings where electricity is made
radioactive (*adjective*)	sending out dangerous rays
reactor (*noun*)	a large structure that makes energy using nuclear power

renewable (*adjective*) able to be replaced

solar cell (*noun*) a device that makes electricity from light and heat from the Sun

static charges (*noun*) electricity made by electrons moving from one material to another

sustainable (*adjective*) using materials and methods to make sure a resource is available in the future

watts (*noun*) units used to measure electrical power

Index